THE HISTORY OF ALCOHOL IN THE FAR EAST

-CHINA, JAPAN, PHILIPPINES, ISLANDS OF THE PACIFIC-

BY

EDWARD RANDOLPH EMERSON

Copyright © 2013 Read Books Ltd.
This book is copyright and may not be
reproduced or copied in any way without
the express permission of the publisher in writing

British Library Cataloguing-in-Publication Data
A catalogue record for this book is available from the
British Library

Contents

WINEMAKING 1
CHAPTER 1 5

Winemaking

The science of wine and winemaking is known as 'oenology', and winemaking, or 'vinification', is the production of wine, starting with selection of the grapes or other produce and ending with bottling the finished product. Although most wine is made from grapes, it may also be made from other fruits, vegetables or plants. Mead, for example, is a wine that is made with honey being the primary ingredient after water and sometimes grain mash, flavoured with spices, fruit or hops dependent on local traditions. Potato wine, rice wine and rhubarb wines are also popular varieties. However, grapes are by far the most common ingredient.

First cultivated in the Near East, the grapevine and the alcoholic beverage produced from fermenting its juice were important to Mesopotamia, Israel, and Egypt and essential aspects of Phoenician, Greek, and Roman civilization. Many of the major wine-producing regions of Western Europe and the Mediterranean were first established during antiquity as great plantations, and it was the Romans who really refined the winemaking process.

Today, wine usually goes through a double process of fermentation. After the grapes are harvested, they are prepared for primary fermentation in a winery, and it is at this stage that red wine making diverges from white wine making. Red wine is made from the must (pulp) of red or black grapes and fermentation occurs together with the grape skins, which give the wine its colour. White wine is made by fermenting juice which is made by pressing crushed grapes to extract a juice; the skins are removed and play no further role. Occasionally white wine is made from red grapes; this is done by extracting their juice with minimal contact with the grapes' skins. Rosé wines are either made from red grapes where the juice is allowed to stay in contact with the dark skins long enough to pick up a pinkish colour (blanc de noir) or by blending red wine and white wine.

In order to embark on the primary fermentation process, yeast may be added to the must for red wine or may occur naturally as ambient yeast on the grapes or in the air. During this fermentation, which often takes between one and two weeks, the yeast converts most of the sugars in the grape juice into ethanol (alcohol) and carbon dioxide. The next process in the making of red wine is secondary fermentation. This is a bacterial fermentation which converts malic acid to lactic acid, thereby decreasing the acid in the wine and softening the taste. Red (and sometimes White) wine is sometimes transferred to oak barrels to mature for a period of weeks or months; a practice which imparts oak aromas to the wine. The end product has been both revered as a highly desirous and delicious status symbol, as well as a mass-produced, cheap form of alcohol.

Interestingly, the altered consciousness produced by wine has been considered *religious* since its origin. The Greeks worshipped Dionysus, the god of winemaking (as well as ritual madness and ecstasy!) and the Romans carried on his cult under the name of Bacchus. Consumption of ritual wine has been a part of Jewish practice since Biblical times and, as part of the Eucharist commemorating Jesus' Last Supper, became even more essential to the Christian Church.

Its importance in the current day, for imbibing, cooking, social and religious purposes, continues. Winemaking itself, especially that on a smaller scale is also experiencing a renaissance, with farmers and individuals alike re-discovering its joy.

CHAPTER 1

In China, a country which has preserved its civil polity for so many thousand years, the art of distillation was known far beyond the date of any of its authentic records. The period of its introduction into that country, in common with the rise and progress of other chemical arts, is, however, concealed amidst the darkness of ages: But taking dates as we find them, sanctioned by respectable authority, and leaving the assumed antiquity of the nation as a point for the discussion of chronologists, we are certainly led to attribute to the people of this empire the merit of an invention which seems to have eluded the grasp of the human intellect in the rest of Asia, Africa, and Europe, until a more advanced period in the history of the world.

There is no doubt whatever, that from the earliest ages the Chinese were acquainted with many of those useful and ingenious preparations, which are still considered as indispensable in the practice of the arts and manufactures of every civilized country. Their knowledge of gunpowder, before it was discovered in Europe, seems to be a fact undisputed, and appears coeval with that of their most distant historic events. Shut up within the bosom of a country, yielding in abundance all the necessaries and even luxuries of life, and satisfied with the articles which it afforded, they felt no desire to seek or encourage an intercourse with foreign nations.* Their inventions, therefore, appear to be entirely their own; the annals of the empire, in the language of Staunton, bear testimony to the fact, and it is confirmed by the consideration of the natural progress of those inventions, and of the state of the Chinese arts at this time.† That they were versed in all the secrets of alchymy, or rather in that branch of it which had for its object a universal *panacea*, long before this fancy engaged the speculations of European practitioners, we have abundant proof‡, since some of their empiricks have from an early period boasted of a specific among their drugs, which insures an immortality like that conferred on Godwin's St. Leon.§ The search after this *elixir vitæ* originated, it appears, among the disciples of the philosopher Lao-kiun, who flourished six hundred years before Christ. Not content with the tranquillity of mind which that teacher of wisdom endeavoured to inculcate, and considering death as too great a barrier to its attainment, they betook themselves to chemistry, and after the labour of ages in a vain endeavour to prevent the dissolution of our species, and after the destruction of three of their emperors, who fell victims to the immortal draught*, they, like the alchemists of Europe, ended their researches under the pretence of discoveries which were never made, and of remedies which could only be administered under all the extravagancies of magic. Indeed, in any country where medicine has not been established, as a regular study, it can scarcely be expected, that the profession of a chemist could be supported with dignity or respectability. But, whether to this search, or to other circumstances, the early knowledge of the Chinese in distillation is to be ascribed, it would be no easy matter to determine. Their acquirements in medicine are so limited, that Navarette says, the greatest part of their physicians are mere farriers; that they know nothing of potions, and that their chief care and skill consists in little more than the recommendation and observance of a regular diet.*

From this unimproved state of an art so important to human existence, it is clear they owe nothing to foreign or factitious aid; and although it might be urged that the Arabians, at an early period, advanced as far as Canton, where they might have communicated some of the discoveries of their physicians and philosophers, it ought to be recollected that it was the spirit of commerce which carried the Mussulmans to the confines of this remote region†, and that the power of the still, if known to them at that time, was altogether applied to the improvement and advancement of medical knowledge; a use, to which, as far as we can learn, it has never yet been devoted in China. The love of literature, as learning leads to the highest posts of honour, has long prevailed amongst this people; and their progress in moral philosophy and the belles lettres has been by no means inconsiderable. To this advancement the knowledge of printing has greatly contributed; but although that art, according to Trigaucius and others, has been known to them above 1776 years, it has remained comparatively stationary, doubtless from the nature of the language, which one would think must render the printing of books troublesome and tedious. The printer or rather engraver of a book has to trace the characters of each leaf on a piece of plank or block of hard wood*, so that for a work of any extent a store of some magnitude is required. What must have been the room requisite for the materials of one of their dictionaries, consisting of 120 large volumes, or for the ancient and modern laws of the country, which the emperor *Tay-tsu*† ordered to be printed in 1380, in 300 volumes! But, notwithstanding this apparent difficulty, books are said to be numerous.* The emperor *Tay-Tsong* is represented to have had a library of 80,000 volumes, the composition of native authors, which was neatly distributed in three large rooms richly adorned; and so addicted to reading was that monarch, that he daily turned over one or two volumes himself†; and the famous library of Ywen-ti, which was burned in *552*, consisted of 140,000 volumes.‡

The whole nation, says a Jesuit missionary, who had a good opportunity of observation, is much addicted to learning and inclined to reading. In one province, we are told§, there are sometimes upwards of 10,000 licentiates and bachelors, and the number of candidates for degrees at a moderate computation amounts to 2,000,000. In the southern provinces of the empire there is scarce a Chinese that cannot read and write.|| I have met, says Navarette, men on the road in sedans, or palankins on men's shoulders, with a book in their hands. In cities I have often seen Mandarins occupied in the same manner; and to induce their children to learn, the tradesmen and shop-keepers might be seen sitting behind their counters with books before them. For the encouragement of students, says the same writer, the example is related of a poor young man who herded cows, and rode upon one of them, as is usual in the country; his love of study was such that he kept a book placed on her horns in such a manner that it served him as a desk, and enabled him to read all the day, by which means he attained to a high station in the state. Another instance is mentioned of a youth, who, being so poor that he could not buy oil for his lamp, studied at night by the light of the moon and the stars: his erudition procured him equally honourable advancement. But, although the application of the Chinese has been sufficiently diligent and laborious, we have no account of any of their publications on the useful or speculative arts. To this circumstance, combined with a constant jealousy and fear of imparting to others a knowledge of these inventions which they consider purely their own, are, perhaps, to be attributed the very brief and unsatisfactory accounts which writers have been able to collect, of the nature and extent of their inebriating beverages. We read that under the government of the emperor *Yu* or *Ta-yu*, before Christ 2207*, the making of ale, or wine, from rice was invented by an ingenious agriculturist named *I-tye**; and that as the use of this liquor was likely to be attended with evil consequences, the emperor expressly forbid the manufacture or drinking of it, under the severest penalties. He even renounced it himself, and dismissed his cup-bearer, lest, as he said, the princes his successors might suffer their hearts to be effeminated with so delicious a beverage.† This, however, had not the desired effect, for having once tasted it, the people could never afterwards entirely abstain from the bewitching draught It was, even at a very early period, carried to such excess, and consumed in such abundance, that, the emperor *Kya*, the Nero of China, in 1836 before Christ, ordered 3000 of his subjects to jump into a large lake which he had prepared and filled with it; while *Chin-vang*, in 1120, thought it prudent to assemble the princes to suppress its manufacture, as the source of infinite misfortune in his dominions.‡ The produce of the grape, it should seem from this, was not so early attended to, although the cultivation of the vine has been known and practised in China from the most remote period. Indeed, all the songs which remain of the early dynasties down to that of *Han*, which commenced 206 years before the Christian era, confirm this opinion, and give us reason to believe that the Chinese have always been fond of wine made from grapes. Grosier says that the emperor *Ouenti*, of the dynasty of *Ouei*, celebrates it with a lyric enthusiasm worthy of Horace or Anacreon; and we find in the large Chinese herbal, book 133, that wine made from grapes was the wine of honour, which several cities presented to their governors and viceroys, and even to the emperor. In 1373, the emperor *Tay-tsu*, who ascended the throne five years before, accepted some of it for the last time from *Tai-yuen*, a city in the province of *Chen-si*, and forbade any more to be presented. "I drink little wine," said the prince; "and I am "unwilling that what I do drink should occasion "any burden to my people." It appears, according to the same writer, that the vine has undergone many revolutions in China. It has never been left out, when orders were issued for rooting up all those trees that encumbered the fields destined for agriculture. The extirpation of the vine has been even carried so far in most of the provinces, during certain reigns, that even the remembrance of it has been entirely forgotten. When it was afterwards allowed to be planted, it would appear, from the manner in which some historians express themselves, that grapes and the vine began then to be known for the first time. This probably has given rise to the opinion that the vine has not been long introduced into China. It is, however, certain, without speaking of remote ages, that the vine and grapes are expressly mentioned in the Chinese annals, under the reign of the emperor *Vou-ty*, who came to the throne in the year 140 before the Christian era; and that, since his time, the use of wine may be traced from dynasty to dynasty, or, as we may say, from reign to reign, even to the fifteenth century. With regard to the present state of the culture of vines in China, Grosier states as a. well known fact, that the two last emperors, with *Kien-long*, who was on the throne when Lord Macartney visited the country, caused a number of new plants to be brought from foreign countries, and that three of the provinces in particular, viz. *Honan, Shan-tong*, and *Shan-si*, have repaired their former losses by the cultivation of them.* Barrow remarked, that in his time no wine was made from the juice of the grape, except by the missionaries near the capital.†

Of rice wine there are different sorts, but none of them have any resemblance to the wines of Europe, either as to taste or quality, being variously compounded, and never allowed in the manufacture to preserve the mere flavour of the original material. That called *mandarin*, being considered of a superior class, is drawn from rice of a particular description, different from that which is eaten.* The grain is steeped for twenty or thirty days in water, and then gently boiled. When it is quite soft and pulpy, and completely diluted and dissolved by the heat, it is allowed a considerable time to ferment, in proper vats prepared for the purpose, generally of glazed earthenware. Several wholesome ingredients are added during the process, mostly simples, and consisting of such fruits and flowers as impart an agreeable flavour and pleasing colour.† At the end of several days, when the motion or agitation occasioned by the fermenting process has subsided, and when the liquor has thrown up all the scum or dross, it is drawn off into glazed vessels, where, by a second species of fermentation, it clears itself, and developes, by the taste and smell, its good or bad qualities. When sufficiently fined, so as to shew, by standing for some time, its body and colour, it is put into small jars, in which way it is commonly sold, and sent through the empire, or to Tonquin and Corea. This wine is usually so strong that it will keep for a great many years, or, as some say, for ages. Within the empire it is principally consumed among the higher orders, who can afford to buy it; and when exported, it sells very dear.* The lees are distilled, and yield a strong and agreeable kind of spirit like brandy; this they call *show-choo, sau-tchoo, sam-tchoo*, (literally burnt) or hot wine. The city of *Kyen-chang*, in the province of *Kyang-li*, is noted for making a fine species of this wine, while that of *Vû-si-hyen*, in *Kyang-nan*, is in great esteem, owing its excellence to the goodness of the water to be found there.† Navarette, in his journey to the imperial residence, remarks, that in the district of the city of *Kian-hoa*, the liquor of this class was made so good that he felt no regret for the wines of Europe. He represents it as exceedingly wholesome, and gives a proof of it, in the instance of a person of rank above seventy years of age with whom he was acquainted, and who had been in the habit of drinking at breakfast, for the greater part of his life, a pint and a half of this wine. Some of the rice wines are so highly perfumed, and so odoriferous, that on opening a bottle, the air of an apartment assumes an agreeable fragrancy; such is the state of perfection to which these people have arrived in the making of this intoxicating luxury.

In many of the provinces an excellent description of wine is made from the palm-tree, called *cha*, a term that is also given, to tea; but the process differs little from that already described as being practised in India, to which we refer the reader. In a country so extensive as China, abounding in all the species of fruit that grow in other parts of the world, such as apples, pears, plums, quinces*, apricots, peaches, figs, pomegranates, mulberries, nectarines, grapes, oranges, lemons, citrous, melons, walnuts, chesnuts, pine-apples, and other fruits peculiar to the soil, with a vast variety of grain and esculent substances, that contain saccharine matter, what, it may be asked, in the hands of so ingenious a people, must be the variety of wines or vinous liquors that daily sparkle on the tables of the luxurious in this remote and secluded region.

The following mode of making beer is observed in China. The liquor is called *tar-asun*, and is extracted from barley or wheat. The grain from which it is produced undergoes a certain degree of malting, after which it is coarsely ground, and put into a kieve, where it is moistened lightly with warm water and closely covered; after it has stood for some time, boiling water is again poured upon it, and the whole is stirred until it appears completely wetted and mixed: this operation being performed, they cover the kieve a third time, letting it stand as before; they then open it again, stirring the whole contents and pouring in boiling water, until the lighter materials rise to the top, and the liquor assumes the strong flavour of grain, &c., which they know by its having gained a deep colour and an adhesive or glutinous consistency. When the liquid has become lukewarm, they pour it into a narrower vessel than the kieve; and after having mixed a small portion of Chinese hops, they put this vessel, with the liquor, down into the earth, for the purpose of fermentation. The Chinese hop is a prepared one, which bears its leaven with itself, and which excites fermentation. As soon as the working has ceased, and the liquor has begun to subside, they fill large bags with it, or rather coarse sacks, made of a suitable thickness for the purpose, which are put into a press. The liquor being extracted, is poured into barrels, which are bunged up with care, and immediately placed in a cellar.* In the distilleries the same process is observed for the preparation of the wort or wash from wheat, rye, or millet, except that no hops are used when the liquor from the grain is intended to be distilled. Before this extract is submitted to any kind of fermentation, it is mixed with a preparation called *pe-ka*, consisting of rice-flour, liquorice-root, anniseed, and garlic; this, it appears, not only accelerates fermentation, but is supposed to impart a peculiar favour. The whole of the mixture being duly fermented, undergoes distillation; and the *Sau-tchoo* thus prepared may, as Barrow remarks, be considered as the basis of the best arrack, which in Java, as already noticed, is exclusively the manufacture of Chinese, and is nothing more than a rectification of the above spirit, with the addition of molasses and juice of the cocoa-nut tree.* Before distillation, the liquor is simply called *tchoo*, or wine; after that the word *show, sau*, or *sam*, is added, to express its hot, burning, or fiery nature. The great materials of distillation throughout all China are rice and millet, the former of which, according to Sir George Staunton, is produced in great abundance in the middle and southern provinces of the empire, while the latter supplies its place in the northern. Of the great extent of the culture of those articles of human food we can scarcely form an idea, even on learning that the mere tribute paid from the different provinces into the royal treasury yearly, as a duty on the lands, amounts in these different kinds of grain to 40,155,490 sacks.† But when the steepest hills and mountains are brought into cultivation, we need scarcely wonder at the agricultural riches of China. The water which runs through the level of the valley is there taught to flow across the mountain, and, from terrace to terrace, to give nourishment to vegetable matter, and assist the hardy labours of the husbandman. One of the missionaries tells us, that in the year 1664 he bought the very best wheat for three ryals (eighteen-pence); and rice of the first quality, "every grain as big as the kernel of a pine-apple," at five ryals (half-a-crown) the bushel. In the province of Shantong*, the same year, wheat was sold for one ryal, or sixpence, the bushel. The nature of this empire being so little liable to change, unless from unfavourable seasons, we are inclined to think that the prices are still the same.† As there does not appear any regulation confining distillation to particular individuals, all the makers of wine distil from the lees, while others manufacture from the grain direct. The produce is distinguished in Europe under the general appellation of *rack, raki*, or *arrack*.‡ The manufacture of this liquor, Grosier tells us, is carried on to a great extent through the whole of the Chinese dominions. Its strength generally exceeds the common proof, and is free from that empyreumatic odour so often perceptible in European spirits. Numbers of carts loaded with it enter Pekin daily. The duty is paid at the gates*, and it is sold publicly in more than a thousand shops that are dispersed throughout the city and suburbs. The sale of this attractive article is conducted in the same way through the whole of the cities, towns, and villages in the fifteen provinces; and it is not a little surprising, that amidst a population of 333,000,000†, the consumption of so dangerous a beverage should be attended with so few fatal consequences, as we are assured on the testimony of some of the most respectable writers‡, that a quarrel or murder occasioned by intoxication is rarely if ever heard of: but we apprehend that to the strictness of the police, and to a regulation rendering every tenth housekeeper accountable for the conduct of the nine neighbouring families§, more than to the steadiness of the Chinese, must be attributed this forbearance, since human nature is much the same in every region of the world. Mr. Abel gives the following picture of the public houses he had an opportunity of visiting while the embassy stopped at the city of *Tung-chow*, on its return from Pekin. These, says he, were large open sheds, fitted up with tables and benches, and affording means of gambling and drinking to the lower orders of the people; they were generally filled with players at dominos or cards, who seemed to enter with intense earnestness into their game. The cards were small pieces of pasteboard, about two inches in length and half an inch in width, having black and red characters painted upon them. The beverage most largely partaken of in these houses was tea, wine, and *sam-su*. All the guests were smoking from pipes of various length, from two to five feet, formed of the young and tender twigs of bamboo, fitted with bowls of white copper about the size of a thimble.* Suoh houses, however, are seldom frequented for the mere love of drinking; and although intoxication is not unfrequent, that vice, it appears, forms no part of the general character of the people.† The rice wines are all drunk warm, for reasons which it would perhaps be difficult to assign, except upon the principle that so powerfully impels the Japanese to follow a similar practice, as noticed in treating of their beverages.*

When scarcity or famine is dreaded, distillation is prohibited, as in Great Britain, by proclamation. Where stills are found afterwards at work, the still-houses are destroyed, the workmen thrown into prison, whipped, and condemned to carry the cangue, a degrading frame of wood, that renders the culprit unable to do any thing for himself as long as he is obliged to wear it. What greatly tends to the encouragement of distillation, is the facility with which coals are procured, and sent by canals through the provinces, while in other parts of the empire less accessible, wood is a substitute of equal value and importance. The skill of the Chinese in distillation is not confined to the manufacture of brandy from rice or millet alone. Besides the quantities that are distilled from the produce of the palm and other fruits, a very ardent spirit, said not to be unworthy of the emperors, is produced from the flesh of sheep.† The nature of the process seems to be as yet a secret to Europeans. Some indeed have stated, that several vegetable substances are employed, but that assertion appears to rest on mere conjecture. The use of this liquor was first introduced by the Tartars, whose fondness for the repasts which the flocks and herds of their native wilds afforded, induced them to subject to the action of the still, the flesh of an animal that had long formed the basis of a more simple, though perhaps not less intoxicating, beverage; we allude to their *lamb-wine**, already mentioned as a favourite drink amongst them. *Kang-hi*, who was of Tartar origin, and wielded the Chinese sceptre for sixty years, encouraged the manufacture of this spirit by the use he made of it himself.† It has, however, never been a favourite in China, and we have little reason to expect that its admirers, should any of them visit Europe, will ever be regaled with a cup of this exhilarating draught. The inhabitants of the province of Quang-tong distil a very pleasant liquor from the flowers of a species of lemon-tree, which are said to possess an exquisite odour, and, like those of the Mahwah or Madhuca of Bahar in India‡, have a strong saccharine quality. The fruit of the tree is almost as big as a man's head; its rind resembles that of the orange, but the substance within is either white or reddish, and has a taste between sweet and sour.* The spirit is perfectly clear and transparent, and is held in high estimation.

From the refuse of their sugar plantations† much rum might be made, but they have not as yet attempted the manufacture of that article. So great is the trade in sugar, that 10,000,000 of lbs. was exported from the country in 1806.‡

The sugar exported from Canton for American consumption in four years, from 1815 to 1819, amounted to 39,670 peculs; and from that port in the same period were exported for European use 21,400 peculs.§ The entire quantity carried from Canton by the American traders from 1804 to 5th January 1819, appears to be 67,663 peculs|| and the quantity imported into Great Britain, the produce of the East Indies and China, for seven years, from 5th January 1815 to 5th

The History of Alcohol in the Far East

January 1821, amounts to 1,073,730 cwt., which, at £2 2s per cwt., gives a sum of £2,254,833, being at the rate of 4 1/2d. per lb.¶ The Chinese are expert in the manufacture of sugar and sugar-candy; the latter has been celebrated. So far back as 1637, both these articles could be purchased for three half-pence per lb., of a quality as white as snow.*

The wines of Europe are now imported into China like other articles of merchandize, and are often sold to considerable advantage. The xeres or sherry wine is preferred on account of its strength, and because it is not liable to change by heat. The Spaniards send wines to Manilla, Macao, &c., from whence the Chinese bring a considerable quantity, especially for the court at Pekin.†

The East India company exclusively exported to China in ten years, from 1810 to 1820, beer alone to the value of £14,309, and wine in bottles and packages for the same period to the amount of £7,383.‡ But the trade has increased, and the following account of all beer, ale, and spirits, both British and Foreign, as well as wine, exported from Great Britain to the East Indies and China for a period of seven years, is inserted here, for the purpose of shewing at one view the extent of this commerce, and its importance as a source of wealth and consumption to our home and foreign manufacture.§

The History of Alcohol in the Far East

Year.	Ale and Beer.			Value.			British Spirits.	Value.			Foreign Spirits.	Value.			Wine.	Value.		
	Tuns.	Hhd.	Gal.	£	s.	d.	Galls.	£	s.	d.	Gallons.	£	s.	d.	Gallons.	£	s.	d.
1814	2,699	1	9	50,021	16	1	5,033	2,938	15	0	207,681	44,016	0	0	360,904	100,251	0	0
1815	5,511	1	24	117,057	7	10	6,981	4,120	14	0	164,885	31,003	0	0	236,271	65,631	0	0
1816	6,821	2	16	137,781	2	10	4,400	3,268	18	5	133,042	24,422	0	0	266,872	74,131	0	0
1817	4,780	0	6	111,187	14	6	3,430	2,815	19	0	124,752	22,718	0	0	198,672	55,187	0	0
1818	3,355	2	19	79,541	1	9	4,075½	3,191	5	0	170,079	38,949	9	9	205,956	57,210	0	0
1819	1,814	2	57	40,398	8	10	1,889	1,406	5	6	215,931	51,633	15	0	178,019	49,449	14	5
1820	3,353	2	3	71,016	8	0	2,740	1,725	16	0	278,533	69,035	18	6	154,606	42,946	2	2

The Americans, also, are carriers of these articles. In the year ending 5th'January 1819, one thousand gallons of gin were imported by them into Canton.* The superior quality of European spirits renders their importation desirable, as much confusion and danger have arisen in the immoderate use made of the ardent spirits of the country by the British, sailors who frequent this port, and of whose habits the Chinese take advantage, by mixing their liquors with ingredients of an irritating and mad-: dening effect. It superinduces a state of inebriety, more ferocious than that occasioned by any other spirit, and leading the men into the most riotous excesses, tends to establish in the minds of the peaceable inhabitants the most unfavourable opinion of the English character.†

In the island of Formosa, which is situated in the Chinese sea, the inhabitants, particularly on the coasts, manufacture rice-wine, and distil a spirit from it, much in the manner already described; but the people of the interior, who are less civilized, make, their drink in a very different way. Like their neighbours, they plant rice and live on the produce; but as they have no wine or other strong liquor, they make in lieu of it another sort of beverage, which, if we may believe Georgius Candidius, a missionary who resided amongst them for a length of time, is very pleasant, and no less strong than other wine. This liquor is made by the women in the following manner; they take a quantity of rice, and boil it until it becomes soft, they then bruise it into a sort of paste, afterwards they take rice flour, which they chew, and put with their saliva into a vessel by itself, till they have a good quantity of it; this they use instead of leaven or yeast, and mixing it among the rice paste, work it together like baker's dough: they put the whole into a large vessel, and after having poured water upon it, let it stand in that state for two months; in the meantime the liquor wterks up like new wine, and the longer it is kept, the better it becomes, and, as it is said, will keep good for many years. It is an agreeable liquor, as clear as pure water at top, but very muddy and thick towards the bottom. The latter, if water be not, as in some instances, added, is frequently eaten with spoons. When they go to work in the fields, they take some of the thick or muddy part along with them, in a vessel of cane, and in another some fresh water ; these two they blend, and when the mixture has stood a while, it serves to refresh them during the heat and labour of the day.*

In the tributary states of Cochin-China and Tonquin, the Chinese mode of making rice-wine and distilling from that grain is very similar to that already described. Sugar of an excellent quality is made in the former country, and the refining of it is carried on there to a degree of perfection unknown, perhaps, in other parts of the world.* The trade in this article is immense j the Chinese alone are said to take 800,000 quintals yearly. It is, however, strange that the inhabitants do not manufacture rum. Grapes are produced in abundance, but they do not appear to be used for wine. From the periodical rains, and consequent inundations, Cochin-China is remarkably fruitful in rice. In all the provinces there are great granaries filled with it, in some of which, it is kept in good preservation upwards of thirty years.† In distilling from this grain the Cochin-Chinese are not inferior to any other Eastern nation: their arrack is the chief and favourite drink; and "they have it in such "plenty," says Barri‡, "that all people in general "drink as much as they will, and become as drunk "as people among us with wine. Graver persons," he adds, "mix that liquor with some other water "distilled from *calamba*, which gives it a delicious "smell, and forms a delicate composition." Lord Macartney and the gentlemen of his suite were regaled with a portion of this spirit at an entertainment given by the governor of the town of Turon, while the ships were anchored in the bay. It was served in small cups, and resembled, in Staunton's opinion, Irish whiskey. The host on that occasion, by way of setting a good example to his guests, filled his cup to the brim, in a true European style of joviality, and after drinking, turned it up, to shew that he had emptied it to the bottom.*

As the kingdoms of Tonquin and Cochin-China were at one period governed by the same laws, there still exists an affinity in the manners, customs, arts, and sciences of the inhabitants. A reciprocity of habit prevails, and we do not find that the Ton-quinese are acquainted with the making of any beverage with which their neighbours are not familiar. The fertility of the country and temperate nature of the climate are said to enable them to grow a great variety of grain. Besides the rice common to the rest of India, they cultivate, according to Grosier, five other kinds peculiar to the soil. The wine from these appears to be excellent; and their arrack, of which they distil large quantities, is much esteemed throughout the East. From the palm, which is abundant, they extract toddy; but it is reckoned by Barron to be bad for the nerves.* The sugar-cane abounds, but we have not heard that any rum is made. The wine is drank warm, as in China, and much of it is used at their religious sacrifices. On these occasions a strange custom prevails, of trying the animals intended as offerings, by pouring warm wine into their ears: if they shake their heads, they are judged proper to be sacrificed; but if they make no motion, they are rejected. The Tonquinese are of a social disposition;

but too much form and ceremony is observed in their visits and entertainments to render them agreeable to strangers. Father Horta† saw a card of invitation for dinner couched in the following terms: "Chao-ting has prepared a repast "of some herbs, cleaned his glasses, and arranged "his house in order, that Se-tong may come and "recreate him with the charms of his conversation "and the eloquence of his learning; he therefore "begs that he will not deny him that divine plea-"sure." When all the persons invited on such occasions are assembled, and before the entertainment begins, the master of the house takes a cup of gold or silver, filled with wine, either of the country or the mandarin of China, and proceeding to the outer court, with his face turned towards the south, pours it out as a peace-offering or libation to the tutelary spirit of his dwelling, This ceremony being over, the guests approach the tables, and before they commence eating, waste an hour in complimenting each other. The person of the greatest distinction in company drinks first. The cups employed to hold the liquor are very small, being scarcely deeper than the shell of a walnut; these however are often replenished, which makes amends for their diminutive size. If the guests chance to play at small games, the losing person is condemned to drink freely, as a forfeit for his ill luck. The arrack and wine of Tonquin are sold every where through the country, and in the public markets held every fifth day.*

The History of Alcohol in the Far East

The Coreans, an ingenious and enterprising people, who inhabit that extensive peninsula washed by the sea of Japan and lying to the northeast of the Chinese empire, manufacture a species of wine, or vinous liquor, from a grain called paniz, or panicum, supposed to be a coarse kind of rice.† This is the only beverage made in the country of which we have any account; but their subjection to the Chinese,. and their consequent intercourse with that people, have no doubt given the Coreans a knowledge of all the liquors peculiar to them, as it is probable, since the more southern and fertile districts of Corea afford wheat, millet, barley, rice, and a variety of fruits to exercise the ingenuity of the inhabitants in the preparation of luxurious potions for the table.

Of Japan, as of the other distant and oriental nations, the early history is but little known. Marco Polo, in the third book of his account of eastern countries, imperfectly describes it under the name of Zipangri. The Portuguese, about the year 1542, were the first who laid open to Europeans a knowledge of those islands.* The inhabitants, though far advanced in civilization, appeared altogether unacquainted with chemistry as a science. In the practice of several of the useful and ingenious arts they had made astonishing proficiency, and in the manufacture of *sacki*, a strong and wholesome beer procured by fermentation from rice, they were not excelled by any other† people. It still appears to be the beverage in general‡ use. Kœmpfer met with it in all the inns at which he stopped on his journey to the metropolis; and although no person whatever is exempt from brewing it, yet there are numbers in the empire who follow no other business than that of making *sacki*. The town of *Muru*, in the province of Bisen, is inhabited chiefly by the brewers of this liquor; and at a village near the city of *Osacca*, it is made to perfection, and in such abundance, that it is sent from thence all over the kingdom, and even exported to other countries by the Dutch and Chinese.* Although *sacki* is drank freely by all descriptions of persons, from the emperor to the meanest subject, its immoderate use is seldom productive of much mischief. Some, indeed, of the lower orders have been known to be beheaded for being drunk and† quarrelsome, but this is of rare occurrence. The beer of Japan, as already remarked, is considered wholesome and pleasant to the taste, but it is of such a nature, that it should be drank not cold, but moderately warm; for when it is not heated, it frequently occasions that dreadful and endemial species of colick, which the Japanese call *senki*, a disease which has proved fatal to many, as well foreigners as natives. To cure this distemper, various means are used, but the principal is the *acupunctura*, or pricking of the abdomen with a needle, so as to let out the hidden or morbifick vapours. "I have been myself," says Kœmpfer, "several times an eye-witness, that in "consequence of these three rows of holes, (for "such are the number of punctures) made accord- "ing to the rules of art, and to a reasonable depth, "the pains of the colick have ceased almost in an "instant, as if they had been charmed‡ away." Independent of the *sacki*, they have a variety of other exhilarating liquors, like the Chinese and other orientals, some made from wheat and other grains. They distil spirits to some extent from rice and wheat; and from the fruits of the country a very nice description of wine is made. Kœmpfer tasted an excellent sort, made from plums, during his stay at Jeddo. They tap the palm, birch, and other trees, from the juices of which they manufacture various beverages with very considerable skill. It has been remarked, that none of these strong liquors are ever tasted by the women; nor even by the men, except on some extraordinary occasions, or on public festivals*; but from the picture which Kœmpfer gives of a large portion of the Japanese females, we are disposed to think they are not quite so abstemious.

Between the island of Kiusiu, the most southerly of that group which forms the empire of Japan, and Formosa, is situated the state of Loo-choo, an island which has lately attracted considerable attention, from the interesting accounts given of it by Surgeon McLeod and Captain Hall, of the Alceste and Lyra. The inhabitants are represented by these gentlemen as possessing most amiable dispositions, and enjoying all the comforts of a land rich in every bounty which nature can bestow. The orange, the lime, the tea-plant, and sugar-cane abound, while rice, wheat, peas, melons, pine apples, &c. are reared in great plenty. Samchoo is distilled by them to considerable perfection, and is made much in the same manner as in China. Nine jars, each containing about fifteen gallons of this liquor, were sent on board the British ships during their stay. Sacki is in use; and of a quality little inferior to that of Japan.*

In some of the neighbouring islands they make a strong drink from the remainder of their crops of corn, rice, fruit, pulse, &c., called *awamuri*.† On the island of Jesso, although the people are but little advanced beyond the state of hunters and fishers, they make a kind of wine resembling *sacki*, which is very strong. This they drink in great quantities, although they are seldom intoxicated; a circumstance ascribed by Father de Angelis, a Jesuit, to their use of the *todo-noovo*, a kind of oil drawn from a fish of the same name, with which they season their rice, and almost all‡ eatables. In the Celebez islands, where the religion is chiefly Mahometan, little or no distillation is carried on, although most of the oriental grains and fruits abound, and the plantain trees are of the best description. On the fruit of this tree the natives in a great measure live, and regale themselves with its inebriating juice, as also on *sagwire*, a very strong species of palm wine.

At Manilla, one of the largest of the Philippine islands, palm trees are said to grow in great perfection, and to exhibit no less than forty species. Such is the magnitude of some of them, that a Jesuit missionary having touched there but a few days, had, through the kindness of a friend, a place prepared for him so capacious, that under two palm leaves alone, he was enabled to say mass, and to sleep secure from the most violent rain. The leaves are shaped like a fan with ridges, and so strong, that no rain however weighty can penetrate them. Here, as well as in Mindoro, another of these islands, a liquor called *tuba* is drawn from them. Jaggory is sometimes made from the juice*; but the sugar-cane is so successfully cultivated, that the manufacture of jaggory is rendered less necessary. Rice is reared with little labour, and in such quantities, as to afford the Chinese, who live on and frequent the island, an opportunity for the exercise of their ingenuity in all the varieties of the brewing process. From the cocoa, nipe, and cabonegro trees, they obtain the materials of an excellent species of brandy.† As Manilla is the great mart and centre of all the Spanish traders in the East, and the several nations with whom they deal, much of the luxuries and comforts of other countries are brought thither. The viceroy lives in great splendour, and at his table, as well as at the table of the higher order of merchants, maybe found most of the wines, spirits, and liquors of Europe, Asia, and Africa.

In casting our eyes over the broad expanse of the Pacific Ocean, successfully traversed by Bougainville, Furneaux, Byron, Wallis, Dampier, Carteret, Cook, La Perouse, and others, we are presented with so many states and islands, that to describe all would be tedious; and as they greatly resemble each other in the productions of the soil, it may be sufficient to give a general idea of the most considerable. The Marian isles, which are about twelve in number, were first discovered by Magellan, who had reason to form so unfavourable an opinion of the inhabitants, that he bestowed on them the name of Ladrones, or *islands of thieves*, in memory of the repeated thefts which he experienced. The people were found to be extremely rude and ignorant, but subsequent navigators have represented them in a more favourable point of view. Wallis, in 1767, remained upon Tinian a month, and seemed pleased with the refreshment he procured. The people speak a language bearing so close a resemblance to that of the Philippine islanders, that they are supposed to have sprung from one common stock; the productions are much the same as in those islands; since the establishment of the Dutch in Guam, one of the principal settlements of this group, the inhabitants have become better acquainted with the enlivening qualities of the cocoa-nut tree, and of the rice cultivated at Rota. De Pages represents the brandy made from the fermented juice of the cocoa tree as excellent. The Manilla ships usually touch at these islands for refreshments, in their voyage from Acapulco. The Carolines, a cluster of islands which lie to the south of the Ladrones, are but little known: they are said to resemble the latter, both as to the natural productions and manners of the people. Captain Wilson, who was wrecked in 1783 upon the coast of Pelew, one of the principal of the group of islands of that name, gives a pleasing picture of the inhabitants. The island is stocked with a great variety of plants, and with trees of various kinds: among them may be ranked the cabbage tree, the bread fruit, and a tree producing a fruit like an almond. Plantains, bananas, Seville oranges and lemons are found. The leaves of the palm serve as thatch for their houses; the milk of cocoa supplies them with drink. A kind of sherbet is made, to which the juice of the orange is added. It is remarkable, that the crews of the ships which were sent from Bombay to these islands in 1790, among the other supplies introduced liquors to the notice of the inhabitants, who thus acquired a taste for the luxurious drinks of their more enlightened visitors. Captain M'Clure, who commanded the ships, remained on this island, resolved to pass the remainder of his life among those ingenious and virtuous people. Of New Britain and New Ireland we know little, but such parts of them as have been explored are considered abundantly fertile. The cocoa and different kinds of palm trees flourish in the forests, while numbers of esculent roots and vegetables are met with in the plains and valleys. The natives are said to be unacquainted with the virtues of the juice of the palm. The Solomon islands seem to be as little known as the two just mentioned, writers being divided as to their number and extent. Quiros, the Spanish navigator who discovered them in 1605, gives a description of the inhabitants, little different from that applicable to other islanders in the Pacific, their arts and habits being much the same. When Cook visited the islands of the New Hebrides, about sixteen in number, in 1774, he found them well wooded and stocked in abundance with sugar-canes and yams. The plantain, cocoas, bananas, and bread-fruit appeared not so abundant as in some of the other islands of this ocean; but, from the fertility of the soil, they might be augmented with very little labour to a supply sufficient for any exigence. Among the inhabitants of the Friendly Islands, the root of the kava-plant is chiefly used as a beverage. It is a species of pepper, esteemed as a valuable article, and cultivated solely for the purpose to which it is applied. In its growth, this plant, of which the natives are very careful, seldom exceeds the height of a man: it has large heart-shaped leaves, and jointed stalks. After it is dug up, it is given to the servants, and in many places to the women, who, breaking it in pieces, scrape the dirt off, while they each chew their portion, which they spit into a piece of plantain leaf. Those who are to prepare the liquor, collect their mouthfuls together, and deposit them in a large wooden bowl, adding a quantity of water proportioned to the degree of strength required. It is then well mixed up with the hands, and wrung hard to make it yield as much liquor as possible. About a quarter of a pint of this beverage is usually put into each cup. On the natives habituated to its use it has no great effect, unless taken to excess; but on strangers it operates like spirits, occasioning intoxication, or rather stupefaction. From the ease with which it is prepared, it may be considered a common beverage; and there is no feast or ceremony held, however trifling, without kava. On one occasion where Cook was present at a funeral ceremony of mourning, a bowl of this drink, containing about a gallon, was prepared; the first cup (which was formed of a plantain leaf), being presented to the king, he ordered it to be given to another person, the second he drank himself, and the third was handed to Captain Cook; cups were then given to the other persons present, until the liquor was exhausted. Each cup, as it was emptied, was thrown upon the ground, whence it was taken up, and carried to be filled again. Scarcely a word was uttered during the whole of this drinking bout, but the utmost gravity was observed by all, from the king to the meanest person present.

The usual drink of almost all these islanders at their meals is water, or cocoa-nut milk, the *kava* being only their morning beverage, unless used as before observed at feasts, and on occasions of ceremony. In the Society Islands this liquor is somewhat differently prepared, and is used among the better sort of people. At Otaheite, the principal of these islands, they pour a small quantity of water upon the root of the ava, and often bake, roast, or bruise the stalks, without chewing, before the infusion. They also bruise the leaves of the plant, and pour water upon them, as upon the root. In large companies it is not drank in the social manner observed among the people at Tongataboo, but from the mode of its preparation it is more intoxicating, and has infinitely worse effects. Its dangerous inebriating quality is exemplified by Captain King in the instance of a man seen by him, who had drank of it to such excess, that he lost his senses, and appeared to be convulsed. Two men held him while in this situation, and busied themselves, by way of a restorative, in plucking off his hair by the roots.* Many of us, says the same writer, who had visited these islands before, were surprized to find, several of the natives, who when we last saw them were remarkable for their size and corpulency, now almost reduced to skeletons; and the cause of this alteration was universally attributed to the use of the *ava*. Their skins were dry, rough, and covered with scales; which they say occasionally fall off, and their skin becomes in some degree renewed. As an excuse for so destructive a practice, they allege its tendency to prevent corpulency; but it enervates them exceedingly, and probably shortens the duration of their lives.

This practice of taking *ava* prevails in the Sandwich Islands, where it is prepared in the same way as among the natives of the Friendly Isles, With this difference, that when a sufficient quantity is collected, they strain it into a Calabash through the fibres of the cocoa nut. The effect it produces is well described by a gentleman who visited them lately.* When a man, says he, first commences taking it, he begins to break out in scales about the head, and it makes the eyes very sore and red; then the neck and breasts, working downwards, till it approaches the feet, when the dose is reduced. At this time the body is covered all over with a white scurf, or scale, resembling the dry scurvy. These scales drop off in the order of their formation, from the head, face, neck, and body, and finally leave a beautiful smooth clear skin, and the frame clear of all disease. The process is also held to be a, certain cure for venereal infection. The writer of this article has known many white men go through a course of this powerful medicine. Women are not allowed to use it; and thus, unhappily, the dreadful disease, first brought to these islands by Captain Cook's crew, remains to curse the inhabitants.

The chiefs, continues the same writer, are much addicted to the use of spirituous liquors, and think nothing of taking a tumbler of strong Jamaica rum at a draught. The higher class of women are, if possible, the greatest drunkards. They distil an excellent spirit from the *tee* root, which grows wild about the mountain, and resembles the beet-root of this country. It is however larger and much sweeter, of a brownish appearance, and in perfection all the year round. The natives collect a quantity of this root, and bake it well under ground; when sufficiently baked, they pound it up in an old canoe kept for that purpose, mixing water with it, and leaving it to ferment for several days. Their stills are formed out of iron pots, which they procure from ships that call there; these they can enlarge to any size, by fixing calabashes, or gourds, with the bottom cut off, and made to fit close on the pot, cemented well with a sort of clay, called peroro; a copper cone is also affixed, with which an old gun barrel is connected, and passes through a calabash of cold water, which cools the spirit. The stills are commonly placed by a stream of water; they take the water when warm out of the cooler, and replace it with cold; by which simple process a spirit is produced not unlike whiskey, only not so strong, but much more pleasant. This spirit is called by the natives *y-wer'a*, which signifies warm water, or luma, in imitation of the word rum. A man of the name of William Stephenson was the first who introduced distilling; he was a convict who had escaped from New South Wales, and lived on the islands for many years; he has left a large family behind him. The credit of first discovering this mode of distilling has been claimed by a person of the name of Young; but, as it has been justly observed, neither of them deserves much praise for the introduction. Mr. Manning, who left Nootka Sound, on the north-west coast of America, at the time when the Spaniards formed an establishment at that place, has cultivated the grape and peach on the island of Woahoo, one of the most important of the Sandwich group; from the former he makes very good wine, and from the latter good peach brandy. In company with this man, the writer, from whom we select this information, went round the island, and found all the plains and vallies in the highest state of cultivation.

In these islands are found very good wheat, rice, Indian corn, and every description of fruit that grows in the West Indies.

* Barrow's Travels in China, 4to. p. 276 and 434, &c.

† Embassy to China, vol. ii. p. 160 and 162, 8vo. edit.

‡ Du Halde, Le Compte, Martini, Osbeck, Grosier, &c. &c.

§ Du Halde.

* *Hyen-Tsong*, in the year 820 of the Christian era, procured some of this liquor, with which it is thought his eunuchs had mixed poison, as he died immediately after drinking it, at the age of forty-three. Du Halde, Annals of the Monarchs, vol. i. p. 200.–*Swen-Tsong*, it appears, had no sooner taken it, in the year 859, than he became a prey to worms, which swarmed in his body, and killed him in a few days, at the age of fifty. Ibid. vol. i. p. 202.–*Shi-Tsong*, or *Kya-Tsing*, also died of this liquor, in 1556, at the age of fifty-eight. Ibid. vol. i. p. 223.–It is said of the emperor *Vu-Ti*, who reigned in China in the year 177 before Christ, that when about to put one of his ministers to death for drinking a cup of this liquor, which had been prepared for himself, he was convinced of his weakness and folly by the following wise and sensible remonstrance of his minister: "If this "drink, Sir, hath made me immortal, how can you put me to "death? but if you can, how does such a frivolous theft deserve "it?" Du Halde, vol. i. p. 177.

* Vide Navarette's Account of China. Barrow's Travels in China, 4to. p. 344. Abel's Journey. Staunton, &c.

† Robertson's Historical Disquisition concerning India, 12mo. pp. 92, 93.

* Pear-tree is mostly used. Mr. Abel, who visited China in the train of Lord Amherst's embassy to the court of Pekin, in 1816 and 1817, says, "Nothing could be more simple than the "method of printing which I have seen practised. On a piece of "wood about two feet square, carved into the necessary charac-"ters, and covered with ink, a thin paper was laid, which, being "pressed down by the hand, received the desired impression. "The use of moveable types in wood is confined to the printing "of the Pekin Gazette, and a few other periodical works. All "others are printed in stereotype. The use of moveable metallic "types may, perhaps, at no distant period, become general in "the empire, as a manufactory of them in block tin is already "established at Macao for the use of the British factory. The "founders and cutters are Chinese, who execute their work with "great precision and dispatch." See Abel's Narrative of a Journey in China, 4to. p. 229.

† Du Halde, vol. i. p. 218, Reign of *Tay-tsu*. It was a whole age before this work appeared.

* Books are printed only on one side, and stitched in thin white paper: their size answers generally to that of our royal octavo. Vide Osbeck's Voyage to China, 8vo. vol. i. p. 277.

† Du Halde, Annals of the Monarchs, vol. i. p. 207. This emperor lived in 982.

‡ Ibid. Reign of *Ywen-ti*, vol. i. p.192.

§ Ibid. vol. i. p. 394.

|| Ibid.

* Du Halde, vol. i. p. 145.

* Might not this be one of the immediate descendants of Noah? Doctor Hales, in his Analysis of Chronology, is of opinion that it was the family of Shem that peopled China; but the writers of the Universal History think that Noah himself, being discontented with the party that had been formed to build the tower of Babel, separated from the main body, and, with some followers, travelling eastwards, at last entered China, and laid the foundation of that vast empire.

† Du Halde, vol. i. p. 433.

‡ Ibid. p. 150. 159.

* Grosier's Description of China, vol. i. chap. v.

† Barrow's Travels in China, 4to. p. 304.

* Du Halde, vol.i. p. 303.

† The Chinese rice wines are in general of a *yellow, red, white,* or *pale* colour.

* Richards' Hist. Tonquin. Du Halde, vol. i. p. 303.

† Ibid.

* Navarette describes the wine made in China from this fruit as of a very delicate and superior kind.

* Vide French Dictionary of Arts and Sciences.

* Barrow, p. 304.

† According to the Chinese geography, Daisin-y-tundshi, the tribute of wheat, in Chinese dân, or bushels, amounts to 6,396,286. The dân is equal to 12,070 cubic inches French. Vide Neuhoff, Embassy, Staunton, &c.

* The wheat sent to the treasury yearly from this province is upwards of 1,271,494 dâns.

† When Barrow was at Pekin, rice sold from 1 1/2*d*. to 2*d*. per lb., bread 4*d*., and wheat-flour from 2 1/2*d*. to 3*d*.

‡ The word arrack, according to Osbeck, appears to take its name from the areca-nut. The Portuguese call this tree Araquero.

* These are nine in number, three on the south front, and two on the other three sides. Vide Grosier, vol. i pp. 396 and 7.

† Respecting the population of China, writers seem not to agree. Lord Macartney and Staunton rate it as above, while the Abbe Grosier makes it 200 millions, and Father Allerstein 198,213,713; others, again, make it only 150 millions. It is, however, generally agreed, that "there seems to be no other "bounds to Chinese populousness than those which the necessity "of subsistence may put to it"

‡ De Guignes, Barrow, Osbeck, &c. &c.

§ Staunton's Embassy, vol. ii. p. 56.

* Abel's Narrative, p. 117.

† Osbeck, vol. i. p. 234, &c.

* Martini and Navarette observe that the Chinese mostly take their wines very hot; and as they like their flavour, they sometimes drink to excess, although they are the reverse of a drunken people. Intoxication, when it occurs, is not considered as a shame, but treated as a jest. The cups used for the drinking of wine are generally made of silver, porcelain, or precious wood, and are of a small size.

† Du Halde, vol. i. p. 303.

* See page 69 of this essay. The Chinese term for this wine is *Kau-yang-tsyew*. It is said to be a very strong and nutritive beverage, and the Tartars delight to get drunk with it. Grosier; vol. ii p. 319. Du Halde, vol. ii. p. 256.

† Grosier, vol. ii. p. 319. Du Halde, vol. i. p. 303.

‡ Ibid. p. 109.

* Du Halde, vol. i. p. 109.

† The sugar-cane grows to great perfection in the southern provinces of the empire.

‡ Malte Brun's Geo. vol. ii. p. 605.

§ Parliamentary Report, 7th May 1821, p. 183.

† Ibid. p. 315.

¶ Ibid. p. 353 to 371.

* Parliamentary Report, 7th May 1821, p. 353 to 371.

† Osbeck's Voy. to China, vol. i. pp. 315 and 16.

‡ Parliamentary Report, 7th May 1821, p.330.

§ Ibid. p. 322, &c.

* Parliamentary Report, 7th May 1821, p. 71.

† Ibid. p. 312.

* Vide Candidius's Account of the Island of Formoss, apud churchill vol. i. p. 405.

* The curious reader may find this described at large in Staunton's Embassy, vol. i. p. 258.

† This preservation of the grain is not uncommon; for we find that corn is kept in the *matamores*, subterraneous vaults, or holes made in the form of a cone, in some of the Barbary states, for thirty years or more. These vaults or holes are closed at the opening, and atmospheric air carefully excluded. *Jackson's Account of Morocco*, p. 102.

‡ *Barri's Account of Cochin-China, Churchill's Coll.* vol. ii.

* Staunton's Embassy, vol. i. p. 255.

* Barron's Description of the Kingdom of Tonquin.

† Letter of Horta, quoted by Grosier.

* Staunton, Grosier, &c.

† Mod. Univ. Hist. vol. vii. p. 329. Malte Brun, vol. ii. p. 498. P. Regis' Geog. Observ. in Du Halde, vol. ii. p. 376, &c.

* Kœmpfer, Introd. Hist. Japan, vol.i. p. 32. Thunberg.

† Kœmpfer, vol. i. p. 121.

‡ Titsingh's Account of Japan.

* Kœmpfer, vol. ii. b. v. p. 426. 469. and 477.

† Ibid. p. 567.

‡ Ibid. Appendix to Hist. of Japan, p. 33.

* Mod. Univ. Hist. vol. vii. p. 376.

* Captain Hall's Account of Loo-choo.

† Mod. Univ. Hist. vol. vii. p. 993.

‡ De Angelis apud Charlevoix. Hist. Japan. Modern Univ. Hist. vol. vii. p. 442.

* Navarette apud Churchill.

† De Page's Trav. vol. i. p. 203.

* Cook's Voy. vol. i. p. 350.

* Literary Gazette, Nov. 1821.

www.ingramcontent.com/pod-product-compliance
Lightning Source LLC
Chambersburg PA
CBHW031226090426
42740CB00007B/731